PowerKids Readers:
Nature Books™

Mountains

Jacqueline Dwyer

The Rosen Publishing Group's
PowerKids Press™
New York

1

To Tom

Published in 2001 by The Rosen Publishing Group, Inc.
29 East 21st Street, New York, NY 10010

First Edition

Book Design: Michael de Guzman
Layout: Felicity Erwin and Edwin Yoo

Photo Credits: p. 1 © F.P.G./Gary Randall; p. 5 © International Stock/Orion; p. 7 © F.P.G./Lee Foster; p. 9 © CORBIS/Galen Rowell; p. 11 © CORBIS/Kennen Ward; p. 13 © F.P.G./John Terence Turner; p. 15 © F.P.G./Charles Benes; p. 17 © CORBIS/W. Wayne Lockwood, M.D.; p. 19 © F.P.G./Sanford Schulwolf; p. 21 © CORBIS/Karl Weatherly; p. 22 (volcano) © International Stock/Warren Faidley.

Dwyer, Jackie, 1970–
   Mountains / by Jacqueline Dwyer. — 1st ed.
     p.cm. — (PowerKids readers. Nature books)
   Summary: Describes different kinds of mountains, some of the animals that live on them, and how people use mountains in various ways.
   ISBN: 978-1-4358-3670-9
     1. Mountains—Juvenile literature. [1. Mountains.] I. Title. II. Series.

GB512 .D98 2000
551.43'2—dc21
                                                              99-055187

Manufactured in the United States of America

2

Contents

Mountains are very high hills.

5

Most mountains are near other mountains. They form a row of mountains. A long row of mountains is called a mountain range.

7

There are mountains all over the world. Mount Everest is the highest mountain in the world.

9

A volcano is a special kind of mountain. Sometimes hot lava pours out of a volcano.

11

It can snow a lot on a mountain. Sometimes there can be too much snow on a mountain. The heavy snow slides down the mountain. This is called an avalanche.

It is windy at the top of a tall mountain. The wind bends the trees that grow near the top of a mountain.

15

There are many animals that live on mountains. Mountain sheep are animals that live on mountains.

Some mountains have snow on them in the summer. People like to hike up snowy mountains. Dogs like to hike up snowy mountains, too.

In the winter, people like to visit the mountains covered in snow. It is fun to ski down a snowy mountain.

Words to Know

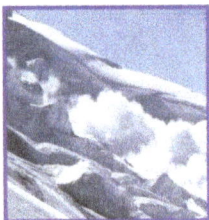
AVALANCHE

HIKE

LAVA

MOUNTAINS

MOUNTAIN SHEEP

SKI

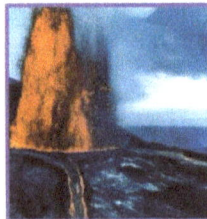
VOLCANO

Here are more books to read about
mountains:
Mountains (Exploring Our World)
by Terry Jennings
Marshall Cavendish

Hills and Mountains (Young Explorer)
by Mark C. W. Sleep
Wayland Publishers Limited

To learn more about mountains, check out
this Web site:
http://encarta.msn.com/find/search.asp?z=
1&pg=1&search=mountains

23

Index

Word Count: 173

Note to Librarians, Teachers, and Parents

PowerKids Readers (Nature Books) are specially designed to help emergent and beginning readers build their skills in reading for information. Simple vocabulary and concepts are paired with photographs of real kids in real-life situations or stunning, detailed images from the natural world around them. Readers will respond to written language by linking meaning with their own everyday experiences and observations. Sentences are short and simple, employing a basic vocabulary of sight words, as well as new words that describe objects or processes that take place in the natural world. Large type, clean design, and photographs corresponding directly to the text all help children to decipher meaning. Features such as a contents page, picture glossary, and index help children get the most out of PowerKids Readers. They also introduce children to the basic elements of a book, which they will encounter in their future reading experiences. Lists of related books and Web sites encourage kids to explore other sources and to continue the process of learning.

www.ingramcontent.com/pod-product-compliance
Lightning Source LLC
Chambersburg PA
CBHW080021240326
41598CB00075B/769